FRIENDS OF THE RICE FARMER

Helpful Insects, Spiders, and Pathogens

B.M. Shepard, A.T. Barrion, and
J.A. Litsinger

W9-BJP-093

1987
International Rice Research Institute
Los Baños, Laguna, Philippines
P.O. Box 933, Manila, Philippines

Contents

3

Foreword

This booklet illustrates representative examples of some of the more common species of predators, parasites, and diseases of insect pests of rice. It can be used with the IRRI booklet *Field problems of tropical rice*, which provides information only on pest species.

Before intelligent decisions about pesticide applications can be made, it is necessary to be able to identify which insect species are pests and which are beneficial. The occurrence of beneficial organisms varies depending upon location, time of year, and crop cultural practices. Thus we made no attempt to rank the groups by their relative importance.

Scientific language has been minimized so that the descriptions can be more easily understood. The pictures will provide an easy way of identifying beneficial species and thereby help prevent unnecessary chemical treatments.

Like *Field Problems*, this booklet was designed to facilitate its easy and inexpensive translation and copublication in languages other than English. IRRI does not ask for payment of royalties or payment for translation of IRRI materials published in developing nations. For details, contact the Communication and Publications Department, International Rice Research Institute, P.O. Box 933, Manila, Philippines.

M.S. Swaminathan
Director General

Introduction

There are rich communities of beneficial insects, spiders, and diseases that attack insect pests of rice. The beneficial species often control insect pests, especially in places where use of broad-spectrum pesticides is avoided. Without these beneficial species the insect pests would multiply so quickly that they would completely consume the rice crop.

Pests have high reproductive capacities to offset the naturally high mortality that they face in nature. For example, a brown planthopper female produces many offspring, but because of the action of predators, parasites, and diseases, only about 1 or 2 will survive after one generation. It is not unusual for 98-99% mortality to occur; otherwise, a pest population explosion can be expected.

Natural enemies also have enemies of their own. Parasites and predators each have predators, parasites, and pathogens. Most predators are cannibalistic, a behavior which ensures that, in the absence of prey, some will survive.

The natural balance between insect pests and their natural enemies is often disrupted by indiscriminate use of chemical insecticides. Although insecticides are needed in some cases, they must be used judiciously in order to save these vulnerable natural control agents.

Predators

Predators often are the most important group of biological control organisms in rice; each predator will consume many prey during its lifetime. They are certainly the most conspicuous forms, and are sometimes confused with pests. Predators occur in almost every part of the rice environment. Some, such as certain spiders, lady beetles, and carabid beetles, search the plants for prey such as leafhoppers, planthoppers, moths, and larvae of stem borers and defoliating caterpillars. Spiders prefer moving prey but some may attack insect eggs. Many species of spiders hunt only at night. Others make webs and collect whatever entered the web throughout the day and night.

Many beetles, some predatory grasshoppers, and crickets prefer insect eggs. It is not uncommon to find 80-90% of the eggs of certain insect pests consumed by predators. An adult wolf spider may attack and consume 5-15 brown planthoppers each day. The immature and adult stages of most predators attack insect pests and many prey are required for the development of each predator.

Other predators, such as water bugs live on the surface of the water in the ricefield. When insect pests such as hoppers, small larvae of stem borers, and leaffolders attempt to disperse, many fall on the water surface and are attacked by water bugs and related predators.

Predators tend to be generalist feeders and often attack other beneficial species when other food is scarce. In general, however, predators feed on those species which occur in greatest abundance such as pests. It is important to realize that a few insect pests occurring at levels which cause no economic damage are helpful for they provide food to maintain populations of beneficial species at levels which can prevent damaging pest outbreaks.

It is extremely costly to mass-rear predators for release in ricefields. There are already many predators in each farmer's field. Predators should be conserved by using broad-spectrum insecticides sparingly or by applying insecticides which are selectively toxic to pests but not to predators.

Parasites

Parasites are generally more host specific than predators. That is why, except for the larger, brightly colored species, they are often overlooked. However, their effect on a pest population can be extremely important.

Whereas predators require several prey to complete their development, parasites normally require only one. Parasites lay their eggs either in groups or singly on, in, or near a host. When a parasite egg hatches and the immature parasite develops, the host usually stops feeding and soon dies.

Many species of parasites attack a single pest species. For example, we have collected 18 species of parasites from leaffolders.

Parasites may attack the eggs, larvae, nymphs, pupae, or adults of the host and, in most cases, they become more effective as host abundance increases. Unlike predators, parasites can find their hosts even when host densities are low.

There have been attempts to introduce parasites from one country to another. In rice, however, most attempts have been unsuccessful because there is already a very rich community of parasites that help to keep pest populations at economically insignificant levels.

Mass rearing of parasites for release to ricefields is useful in certain situations, but normally is very expensive and requires organization.

Parasites should be spared by using insecticides judiciously.

Pathogens

Various microorganisms can infect and kill insect pests of rice. The major groups are fungi, viruses, and bacteria. Nematodes and some other organisms also occur.

Fungi are by far the most important on various leafhoppers and planthoppers. It is not uncommon to find outbreaks of the fungi *Hirsutella citriformis, Beauveria bassiana,* or *Metarhizium* spp., which infect and kill 90-95% of a population of brown planthopper.

Viruses and fungi often control caterpillar pests. The most important are the nuclear polyhedrosis and granulosis viruses. Virus-infected caterpillars cease feeding and the body content liquefies. Thus, the bodies become flaccid and often hang from the rice plant. Viruses have been recorded from almost every species of caterpillar pests of rice. We have observed outbreaks of viruses in populations of leaffolders and cutworms.

The most striking disease outbreaks on caterpillar pests are caused by the fungus *Nomuraea rileyi*. We have recorded a high incidence of this fungus in populations of defoliators. In some cases, the caterpillar populations did not reach economically damaging levels because the fungus was present.

Diseases of pests can be mass-produced at a low cost in liquid or powder form that can be sprayed like an ordinary insecticide.

Predators — lady beetles

Micraspis sp.
Micraspis crocea (Mulsant)
Coleoptera: Coccinellidae

Micraspis sp. (Fig. 1) is a typical coccinellid beetle — oval and brightly colored in shades of red. Lady beetles are active during the day in the upper half of the rice canopy in dryland and wetland habitats. Both the adult (Fig. 2) and the dark larva (Fig. 3) of *Micraspis crocea* feed on small planthopper prey as well as on small larvae and exposed eggs. Adults of *M. crocea* are yellow with variable spots behind the head (Fig. 4).

4.1. No spot on hard wing

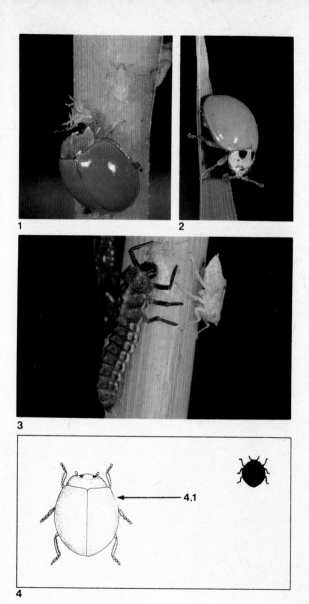

1

2

3

4.1

4

11

5

6

6.1

7

8

8.1

9

Predators — lady beetles

Harmonia octomaculata (Fabricius)
Menochilus sexmaculatus (Fabricius)
Coleoptera: Coccinellidae

Harmonia octomaculata (Fig. 5, 6) and *Menochilus sexmaculatus* (Fig. 7, 8) are black-spotted lady beetles that catch slow-moving prey. Adults are quick to fall off plants or fly when disturbed. A lady beetle takes 1-2 weeks to develop from egg to adult and produces 150-200 offspring in 6-10 weeks. Lady beetle larvae are more voracious than adults and consume 5-10 prey (eggs, nymphs, larvae, adults) daily. A *H. octomaculata* larva is shown feeding on a planthopper nymph (Fig. 9).

6.1. Each hard wing with five spots

8.1. Hard wings with 3 pairs of markings

Predator — ground beetle

Ophionea nigrofasciata (Schmidt-Goebel)
Coleoptera: Carabidae
Ground beetles are active hard-bodied insects. Both the shiny black larvae and reddish brown adults (Fig. 10, 11) actively search the rice canopy for leaffolder larvae. *Ophionea nigrofasciata* can be found within the folded leaf chambers made by leaffolder larvae. The predator larva pupates in the soil of wetland rice bunds or dryland fields. Each voracious predator consumes 3-5 larvae per day, leaving only the head capsules. The adult also preys on planthoppers.

11.1. Bluish black band with two white spots on both ends

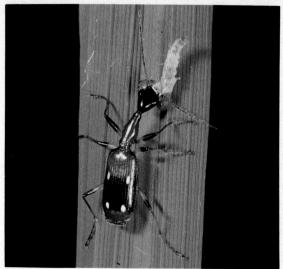

10

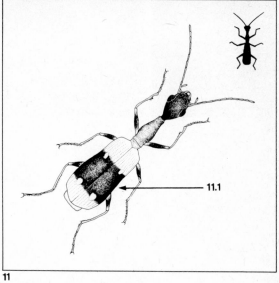

11.1

11

15

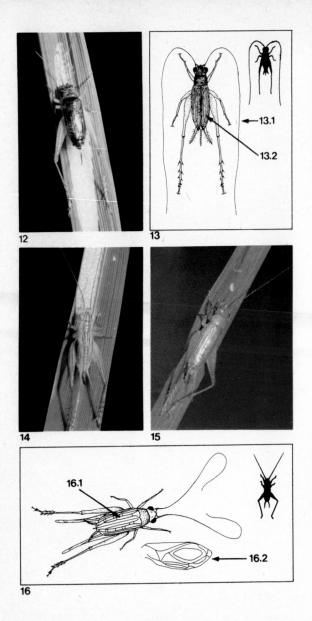

12

13

13.1

13.2

14

15

16.1

16.2

16

16

Predators — crickets

Metioche vittaticollis (Stål)
Anaxipha longipennis (Serville)
Orthoptera: Gryllidae
Sword-tailed crickets occur in wetland and dry-land habitats, and jump from plant to plant when they are disturbed. Most adults lose their hind wings after locating ricefields. Older nymphs have wing pads.

A *Metioche vittaticollis* adult (Fig. 12, 13) is black and nymphs (Fig. 14) are pale with brownish stripes. The adult (Fig. 15, 16) and nymph of *Anaxipha longipennis,* a related egg predator, are brown.

The sword-like ovipositors of *M. vittaticollis* are used to insert eggs into leaf sheaths of rice and grasses. The life cycle from egg to adult lasts 60-80 days and a female will produce 40-80 young. Adults and nymphs are primarily egg predators but will also feed on small larvae and hoppers. They feed on eggs of striped and dark-headed stem borers, leaffolders, armyworms, whorl maggots, and nymphs of planthoppers and leafhoppers.

13.1. Long antenna
13.2. Front wing has a few cross veins

16.1. Front wing has more cross veins
16.2. Male with ringed pattern on front wing

Predator — grasshopper

Conocephalus longipennis (de Haan)
Orthoptera: Tettigoniidae
Meadow grasshoppers are large insects with slanted faces. They can be distinguished from true grasshoppers by their long antennae, which are more than twice as long as their body. Adults are quite active and readily fly when disturbed. They are active at night and are more abundant in older fields. The green nymphs (Fig. 17) can be distinguished from the green and yellow adults (Fig. 18, 19) by the absence of wings and the sword-like ovipositor. The adults live 3-4 months.

Tettigoniids feed on rice foliage and *Conocephalus longipennis* has dual food habits. Aside from feeding on rice leaves and panicles, *Conocephalus* is also a predator of rice bug and stem borer eggs as well as planthopper and leafhopper nymphs. Each predator can consume 3-4 yellow stem borer egg masses a day.

19.1. Slant face
19.2. Long antenna
19.3. Long sword-like ovipositor

17

18

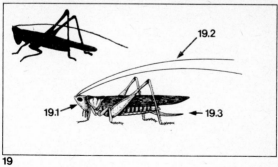

19.2

19.1

19.3

19

20

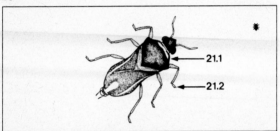

21.1 →
21.2 →

21

22

Predator — water bug

Microvelia douglasi atrolineata Bergroth
Hemiptera: Veliidae

The small fast-moving ripple bugs can be very abundant in flooded fields. Adults and nymphs live on the water surface. The broad-shouldered adults can be either winged or wingless (Fig. 20). The wingless form lacks the black and white markings on the neck and front wings. *Microvelia* (Fig. 21) can be distinguished from other water bugs by its small size and 1-segmented front tarsi. Each female lays 20-30 eggs in rice stems above the water line. The life span is 1-2 months and the winged adults disperse when rice paddies dry up.

Adults congregate to feed on planthopper nymphs which frequently fall onto the water. The nymph (Fig. 22) also feeds on hopper nymphs as well as on other small, soft-bodied insects. *Microvelia* is a more successful predator when attacking in groups and young hopper nymphs are more readily subdued than older and larger prey. Each *Microvelia* can prey on 4-7 hoppers per day.

21.1. Broad shoulder
21.2. 1-segmented front foot (tarsus)

Predator — water bug

Mesovelia vittigera (Horvath)
Hemiptera: Mesoveliidae

Water treaders, like other aquatic bugs, are found only in wetland habitats. The pale green adults are larger than *Microvelia* but are less abundant. Like *Microvelia*, they also have wingless (Fig. 23) and winged (Fig. 24) adults. Wingless adults are more common in ricefields and tend to aggregate near rice bunds. *Mesovelia* adults and nymphs feed mainly on stem borer larvae and hoppers that fall on the water surface. They are solitary feeders.

24.1. Head longer than wide
24.2. Foot (tarsus) with small basal segment

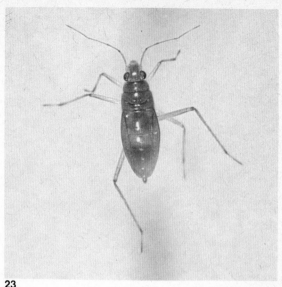

23

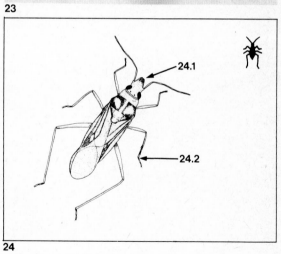

24.1

24.2

24

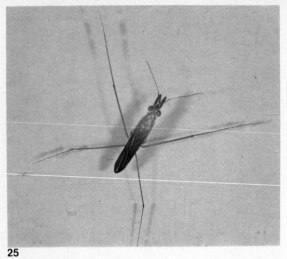

25

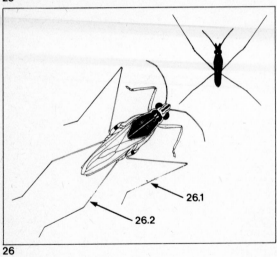

26.1

26.2

26

24

Predator — water bug

Limnogonus fossarum (Fabricius)
Hemiptera: Gerridae

Water striders are large, long-legged, and extremely fast. *Limnogonus fossarum* adults are black with two pairs of very long hind legs (Fig. 25, 26). The middle pair of legs act like oars and project toward the front at rest. Few will be seen in ricefields because they are easily frightened and quick to move away from any disturbance. *Limnogonus* live for 1-1.5 months and lay 10-30 eggs in rice stem above the water line.

Adults and nymphs prey on rice hoppers, moths, and larvae that drop onto the water surface. Each water strider takes 5-10 prey daily.

26.1. Long and slender rear legs
26.2 Hind thigh (femur) beyond end of abdomen

Predator — plant bug

Cyrtorhinus lividipennis Reuter
Hemiptera: Miridae
Cyrtorhinus is an example of a species belonging to a plant-feeding group that secondarily became predators, preferring planthopper and leafhopper eggs and young nymphs. The green and black adults (Fig. 27, 28) and nymphs can become highly abundant in hopper-infested fields, both wetland and dryland. *Cyrtorhinus* eggs are laid in plant tissue, develop to adults in 2-3 weeks, and produce 10-20 young. They search leaf sheaths and stems for hopper eggs, which are sucked dry by piercing mouthparts. Each predator consumes 7-10 eggs or 1-5 hoppers a day.

28.1. Black spot on thorax
28.2. Green membranous wing

27

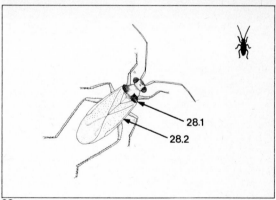

28.1

28.2

28

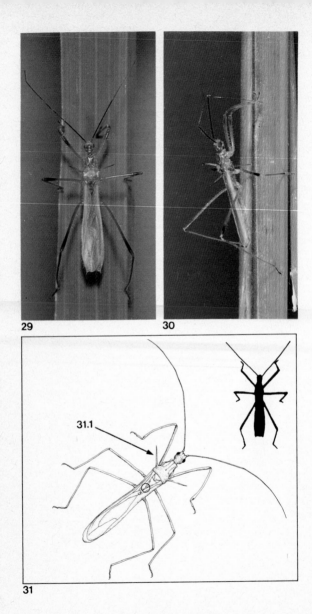

29

30

31.1

31

Predator — plant bug

Polytoxus fuscovittatus (Stål)
Hemiptera: Reduviidae
Assassin bugs as solitary predators rarely become abundant in either the wetlands or drylands. The brown adults of *Polytoxus* have three prominent spines on their back (Fig. 29, 30, 31). Assassin bugs occur in the rice canopy in search of prey, mainly larvae of moths and butterflies. They attack a prey even much larger than themselves, piercing its body with their needle-like mouth-parts to inject a paralyzing toxin.

31.1. Body with three spines

Predators — damselflies

Agriocnemis pygmaea (Rambur)
Agriocnemis femina femina (Brauer)
Odonata: Coenagrionidae

The narrow-winged damselflies are weak fliers compared to their dragonfly cousins. The yellow-green and black adults have a long slender abdomen. Males are more colorful than females. The tip of the abdomen of *A. pygmaea* males (Fig. 32, 33) is orange. The *A. f. femina* (Fig. 34) male has a blue-green abdominal tip and sides of the thorax while the female has a greenish body. Figure 35 shows a mating pair of *A. f. femina* with the female feeding on a leaffolder moth. The male hooks onto the back of the female so they can fly while mating to escape their enemies. Damselfly nymphs are aquatic and can climb up rice stems to search for hopper nymphs. Adults normally fly below the rice canopy searching for flying insects as well as hoppers on plants.

33.1. Tip of abdomen orange
34.1. Tip of abdomen blue-green

32

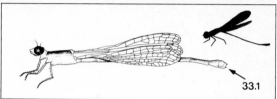

33 33.1

34 34.1

35

36

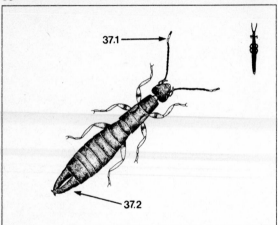

37.1 →

← 37.2

37

32

Predator — earwig

Euborellia stali (Dohrn)
Dermaptera: Carcinophoridae
Earwigs have a characteristic hind pair of forcep-like pinchers that are used for defense rather than for capturing prey. *Euborellia* are shiny black with white bands between the abdominal segments and a white spot on the tip of each antenna (Fig. 36, 37). They are more common in dryland habitats and nest in soil at the base of rice hills. Digging in the soil is the best way to find them. Females show maternal care for the 200-350 eggs that each lays. Adults live 3-5 months and are most active at night. Earwigs enter stem borer tunnels in search of larvae. Occasionally they climb the foliage to prey on leaffolder larvae. They can consume 20-30 prey daily.

37.1. Two white antennal segments
37.2. Forceps

Predator — ant

Solenopsis geminata (Fabricius)
Hymenoptera: Formicidae
Solenopsis (Fig. 38) are fire ants and can inflict a painful bite on the feet and legs of any person walking on a rice bund. These reddish to brown ants make nests in dryland fields as well as in bunds of wetland rice. They normally search for food several meters from their nests. Fire ants are quick to colonize a newly established field, making nests for hundreds and even thousands of workers and soldiers. Ants prey on a wide variety of insects and small animals. They also carry seed from dry ricefields to their nests. Specialized workers break up the seed into edible food for the young. *Solenopsis* will feed on any insect it can subdue. Eggs of black bug (Fig. 39) can be attacked. Even adult black bugs (Fig. 40) are subject to predation by these strong, aggressive insects.

38.1. Head hairy and has 12-segmented antennae
38.2. Segment with two bumps (nodes)

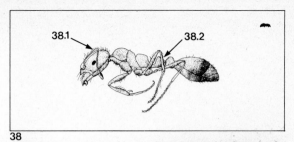

38.1 38.2

38

39

40

35

41

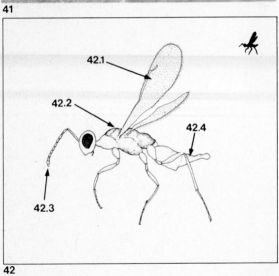

42.1

42.2

42.3

42.4

42

Predator of hopper eggs — wasp

Panstenon nr. *collaris* Boucek
Hymenoptera: Pteromalidae

Pteromalids are small wasps with 5-segmented tarsi and reduced wing veins. *Panstenon* is a small metallic blue-green wasp with reduced wing venation (Fig. 41). It has 13 antennal segments (Fig. 42) while *Tetrastichus*, an eulophid, has only 9. *Panstenon* prefers the wetland rice habitat. The female lays 1 or 2 eggs in a rice tiller. Upon hatching, the small C-shaped parasite larvae search for egg masses of planthoppers and leafhoppers. A larva preys on 4-8 eggs a day. The naked yellowish pupa is found between or within the tiller and the adult wasp emerges in 4-6 days.

42.1. Front wing long
42.2. Coarse middle segment
42.3. Antenna 13-segmented
42.4. Abdomen slender and tapers at end

Predator — wolf spider

Lycosa pseudoannulata (Boesenberg and Strand)
Araneae: Lycosidae

Lycosa pseudoannulata has a fork-shaped mark on the back and the abdomen has white markings. Wolf spiders are highly mobile and readily colonize newly prepared wetland or dryland ricefields. They colonize fields early and prey on pests before the latter increase to damaging levels. The female lays 200-400 eggs in a lifetime of 3-4 months. Eventually 60-80 spiderlings hatch and ride on the back of the female.

Lycosa are most common among tillers at the base of plants and scamper across the water surface when disturbed. They do not make webs but attack their prey directly. Adults feed on a range of insect pests including stem borer moths (Fig. 43, 44). Spiderlings also attack planthopper (Fig. 45, 46) and leafhopper nymphs. Wolf spiders consume 5-15 prey a day. Males (Fig. 47) have enlarged palps.

47.1. Male with enlarged palp
47.2. Fork-shaped light band
47.3. Abdomen with markings

43

44

45

46

47.1
47.2
47.3

47

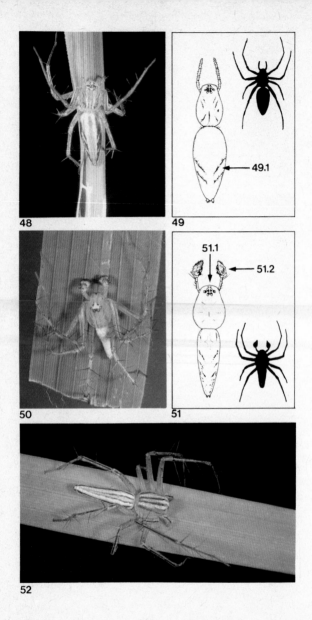

48

49

49.1

51.1

51.2

50

51

52

40

Predators — lynx spiders

Oxyopes javanus Thorell
Oxyopes lineatipes (C.L. Koch)
Araneae: Oxyopidae

Lynx spiders are hunters and build no webs. The *Oxyopes javanus* female has two pairs of diagonal white bands on the sides of the abdomen (Fig. 48, 49) and the male has enlarged palps (Fig. 50, 51). *O. lineatipes* has two reddish brown and two white stripes running along the abdomen (Fig. 52). The female guards its cocoon-like egg mass laid on foliage. These spiders produce 200-350 young and live 3-5 months.

Lynx spiders live within the rice canopy, prefer drier habitats, and colonize ricefields after canopy development. Unlike wolf spiders, they hide from their prey, mostly moths, until within striking distance. They fill an important role, killing 2-3 moths daily and thus preventing a new generation of the pests from building up.

49.1. Diagonal band

51.1. Circular eye pattern
51.2. Enlarged palp

Predator — jumping spider

Phidippus sp.
Araneae: Salticidae

Jumping spiders have two bulging eyes; unlike wolf spiders, they are not quick to move when disturbed. *Phidippus* have brown hairs on the body (Fig. 53). Eggs are laid in an elongated egg mass covered with silk within a folded leaf. The female guards the egg mass and produces 60-90 offspring. *Phidippus* live 2-4 months. They prefer dryland habitats and remain within the rice foliage. They hide in a small retreat web in a folded leaf and fold another leaf within which they lie and wait for leafhopper prey. *Phidippus* prey on green leafhoppers (Fig. 54) and other small insects, and may consume 2-8 a day.

53.1. Large front eyes

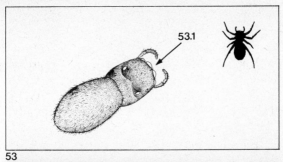

53

54

55

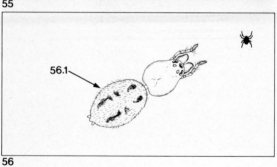

56.1

56

44

Predator — dwarf spider

Atypena (=Callitrichia) formosana (Oi)
Araneae: Linyphiidae

Dwarf spiders are often confused as spiderlings of other species because they are small and up to 30-40 of them can be found at the base of a rice hill. The *Atypena* adult (Fig. 55, 56) has three pairs of gray markings on the back of the abdomen. Globular eggs are laid in masses covered with a thin layer of silk on dried leaf sheaths and receive no maternal care. A female produces 80-100 young. Dwarf spiders prefer wetland habitats and make irregular webs within the base of rice tillers above the water line. They are slow moving and catch most of their prey in webs. They also can hunt directly. *Atypena* live 1.5-2 months and prey on young leafhopper and planthopper nymphs, consuming 4-5 a day.

56.1. Abdomen with three pairs of spots

Predators — orb spiders

Argiope catenulata (Doleschall)
Araneus inustus (L. Koch)
Araneae: Araneidae

Orb weavers are highly colorful and make circular webs in the rice canopy, capturing flying prey as large as butterflies and grasshoppers. They live 2-3 months and lay 600-800 eggs. They are late colonizers of ricefields, but are found in all rice environments.

The female *Argiope catenulata* (Fig. 57) has yellow and grayish white markings on the abdomen; the male is smaller and reddish brown (Fig. 58). The eggs of *Argiope* are inside a light brown cocoon that hangs on the web. When the day is hot, the male and female seek shelter under leaves beside the web; when the day is cloudy, the female waits at the center of the web and the male waits nearby. The more the trapped prey tries to wiggle free, the more it becomes entangled in the strong sticky webs (Fig. 59).

Araneus inustus has a black ovoid band on the abdomen (Fig. 60, 61). The female lays eggs within folded leaves and covers them with white silken threads. *Araneus* prey on small insects — leafhoppers, planthoppers, and flies.

57.1. Abdomen with yellow and grayish white markings
59.1. White stabilimenta (zigzag webbing) on the orb web of *Argiope*
61.1. Black ovoid band on abdomen

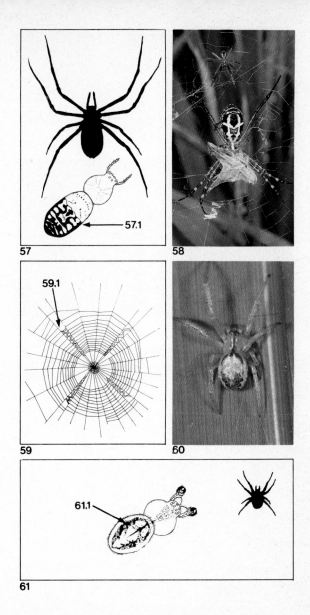

57

58

59.1

59

60

61.1

61

47

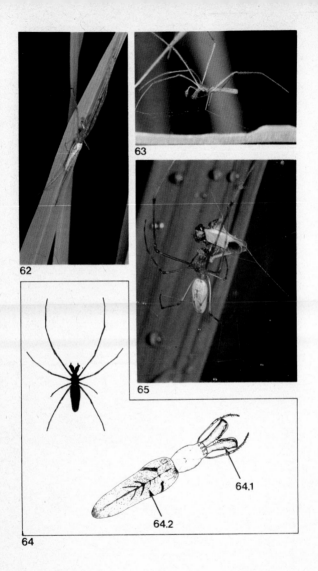

62

63

65

64.1

64.2

64

Predator — long-jawed spider

Tetragnatha maxillosa Thorell
Araneae: Tetragnathidae
Tetragnatha have long legs and bodies and are commonly seen lying outstretched along a rice leaf (Fig. 62). The males can be recognized by the enlarged jaws (Fig. 63, 64). Tetragnathids live 1-3 months and lay 100-200 eggs. The eggs are laid in a mass covered in cottony silk in the upper half of rice plants.

Tetragnathids prefer wetlands where they rest in the rice canopy during midday and wait in their webs in the morning. They spin a ring-shaped web, but the web is weak. When leafhopper prey (Fig. 65), flies, or moths hit the web, the spider quickly wraps them in silk. One tetragnathid kills 2-3 prey daily.

64.1. Long jaw
64.2. Elongated body

Parasite or predator of stem borer eggs — wasp

Tetrastichus schoenobii Ferriere
Hymenoptera: Eulophidae

Tetrastichus adults (Fig. 66) are metallic blue-green. Although barely visible to the naked eye, they are abundant in wetland and dryland rice-fields. Each female wasp produces 10-60 offspring. Several wasps may parasitize an egg mass of yellow or white stem borers (Fig. 67). Prior to oviposition the female examines the egg mass (Fig. 68) in search of areas to probe through the hair mat. When stem borer eggs are located, the abdomen enlarges from the pressure used to force the ovipositor downward (Fig. 69). One egg is laid in each stem borer egg. The parasite larva hatches inside the stem borer egg in 1-2 days . When one host egg is consumed, the larva emerges and preys on other eggs. At least 3 stem borer eggs are needed for development of each wasp. Development from egg to adult stage takes 10-14 days. Parasite wasps emerge 1-2 days later than unparasitized stem borer larvae. *Tetrastichus* will also parasitize *Chilo* stem borer pupae.

66.1. Antenna 8-segmented
66.2. Hairs on wings not in rows
66.3. Foot (tarsus) 4-segmented

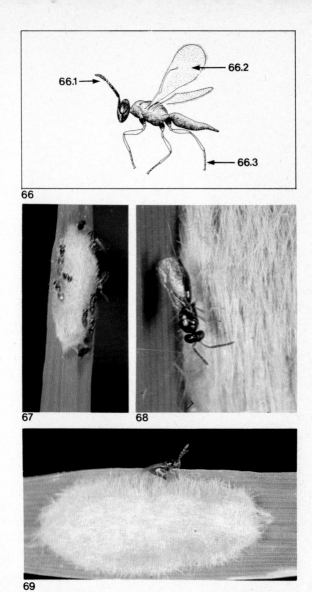

66

67

68

69

51

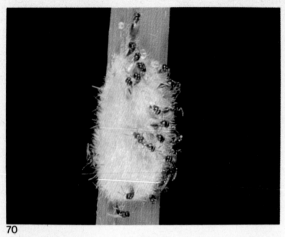

70

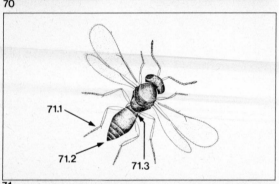

71.1

71.2

71.3

71

72

Parasite of stem borer eggs — wasp

Telenomus rowani (Gahan)
Hymenoptera: Scelionidae

Telenomus is black and barely half the size of *Tetrastichus. T. rowani* (Fig. 70, 71) also attacks eggs of yellow and white stem borers and are as abundant as *Tetrastichus* in wetland and dryland ricefields. Several wasps lay eggs on a single egg mass, but only one *Telenomus* can develop within each egg. Development from egg to adult stage takes 10-14 days and the wasps emerge (Fig. 72) by making characteristic shot holes in the hair mat. The *Telenomus* female seeks out the female moth and attaches itself to the hair at the end of the host's abdomen. The wasp then is carried by the stem borer moth searching for a place in which to lay eggs. *Telenomus* lays its eggs in the stem borer eggs that are being laid and before the latter are covered with body hair. A female parasitizes 20-40 eggs and lives 2-4 days or longer if nectar or sugar solution is provided. Both *Tetrastichus* and *Telenomus* may parasitize the same egg mass but not the same egg.

71.1. Foot (tarsus) 5-segmented
71.2. Pointed abdomen
71.3. Basal abdominal segment with rib-like structures

Parasites of black bug eggs — wasps

Psix lacunatus Johnson and Masner
Telenomus cyrus (Nixon)
Hymenoptera: Scelionidae
Scelionids are egg parasites of moths and bugs. *P. lacunatus* (Fig. 73) and *T. cyrus* (Fig. 74) parasitize the eggs of black bugs. Parasitized eggs are grayish with irregular exit holes; unparasitized eggs are white with egg covers (Fig. 75). *Telenomus* has short hairs on its eyes and a smooth body (Fig. 76); *Psix* has hairless eyes and a pitted body (Fig. 77). *Telenomus* is highly aggressive and will parasitize black bug eggs in spite of the female guarding the egg mass. A wasp that parasitizes an egg leaves a scent that other parasites detect. This odor prevents other wasps from parasitizing the same eggs. Adult wasps live from a few days to a week, or longer.

76.1. Broad second abdominal segment
77.1. Sides of body with many pits

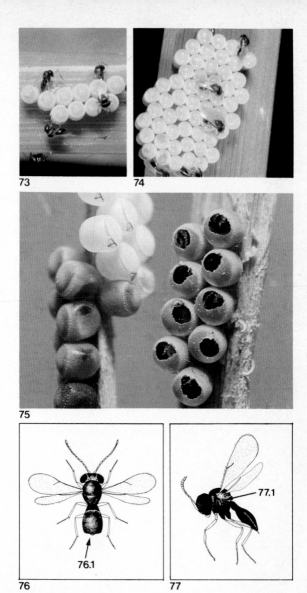

73 74

75

76.1

76

77.1

77

55

78

79

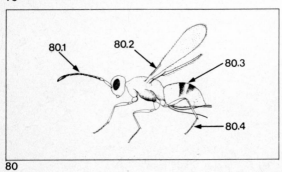

80.1
80.2
80.3
80.4

80

Parasites of hopper eggs — wasps

Gonatocerus spp.
Hymenoptera: Mymaridae

Mymarids are tiny wasps that have oar-shaped wings with long hairs on the margins (Fig. 78). More than five species of *Gonatocerus* parasitize the eggs of leafhoppers and planthoppers. Adults are brown to dark yellow brown with short waists. Males have 13 antennal segments, and females have 11 (Fig. 79, 80). The female can reproduce without mating. She locates host eggs using her antennae. When eggs are located, the wasp raises her body upwards while aiming the ovipositor to place one of her eggs in each host egg. Parasitized eggs turn brownish yellow to reddish yellow while normal eggs are white. Eggs develop into adults in 11-17 days. Wasps live 6-7 days and parasitize an average of 8 eggs a day.

80.1. 8 middle segments
80.2. Marginal vein elongated
80.3. Abdomen arched
80.4. Foot (tarsus) 5-segmented

Parasites of hopper eggs — wasps

Anagrus optabilis (Perkins)
Anagrus flaveolus Waterhouse
Hymenoptera: Mymaridae

Anagrus are solitary egg parasites of leafhoppers and planthoppers. Two species are common in all rice environments. *A. optabilis* (Fig. 81) has a row of hairs on front wings and a long third antennal segment, while *A. flaveolus* (Fig. 82) has 3-4 irregular rows and short third antennal segment. The slender and elongate adults are tiny orange réd to red. Males have 13 antennal segments, but females have 9. Unlike their close relative *Gonatocerus*, *Anagrus* have only 4-segmented tarsi.

Females can reproduce without mating and locate host eggs by drumming the rice stem with their antennae. Once host eggs are detected, the female wasp makes rapid antennal palpations and stretches its legs to allow immediate insertion of the ovipositor into the egg mass. Parasitized eggs turn deep orange red (Fig. 83) while normal eggs are white. Eggs develop into adults in 11-13 days. Wasps live 2-6 days and parasitize 15-30 eggs a day.

81.1. Ovipositor almost at tip of abdomen
81.2. One row of hairs on front wing
81.3. Third antennal segment long

82.1. Ovipositor beyond tip of abdomen
82.2. Front wings with 3-4 irregular rows of hairs
82.3. Third antennal segment short

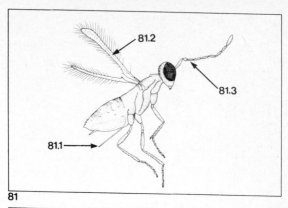

81

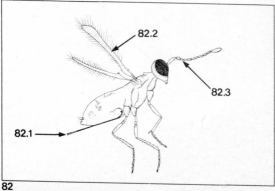

82

83

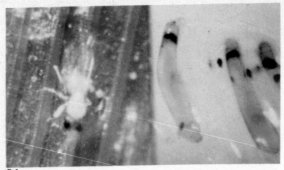

84

85.2

85.1

85.3

85

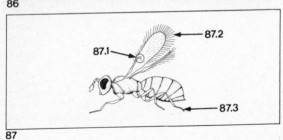

86

87.2

87.1

87.3

87

Parasites of hopper eggs — wasps

Oligosita naias Girault
Oligosita aesopi Girault
Hymenoptera: Trichogrammatidae

Trichogrammatids are minute wasps with 3-segmented tarsi. They are parasitic on eggs of moths and butterflies. *Oligosita*, however, parasitize eggs of leafhoppers and planthoppers. Adults are greenish yellow with transparent wings. Two closely similar species are common in wetland rice environments. *O. naias* (Fig. 84 left, 85) has long hairs on the wing margin and a four-sided wing cell. The females' searching and parasitic behavior is similar to that of *Gonatocerus*. Parasitized eggs are lemon yellow (Fig. 84, right). *O. aesopi* (Fig. 86, 87) has hairs on the margin of the wing shorter than the width of the wing at the triangular cell. Eggs become adults in 11-12 days. Females live 2-5 days and parasitize 2-8 eggs a day.

85.1. Four-sided cell
85.2. Long hairs on wing margin
85.3. Three-segmented foot (tarsus)

87.1. Triangular cell
87.2. Short hairs on wing margin
87.3. Three-segmented foot (tarsus)

Parasite of eggs and pupae — wasp

Trichomalopsis apanteloctena (Crawford)
Hymenoptera: Pteromalidae

Pteromalids are small wasps with 5 tarsal segments, reduced wing venation, and pitted bodies. The adult of *T. apanteloctena* is metallic green with red eyes, yellow legs, and short antennae (Fig. 88, 89). This parasite has been reared from eggs of yellow stem borer and the pupae of rice skippers, green hairy caterpillar, and striped and dark-headed stem borers. It is a secondary parasite of *Goniozus* sp. The wasps hover above dryland rice canopies in search of hosts. One wasp emerges from each egg host, but 20-50 emerge from a parasitized pupa.

89.1. Face nearly rectangular
89.2. Sides of thorax coarse
89.3. Second abdominal segment broad

88

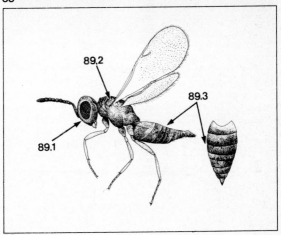

89

90

91

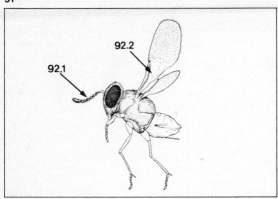

92

Parasite of leaffolder eggs — wasp

Copidosomopsis nacoleiae (Eady)
Hymenoptera: Encyrtidae

Encyrtids are small wasps with an enlarged pair of middle legs used for jumping. The gray to black adult of *Copidosomopsis* is barely visible to the naked eye. Under a microscope, the wings are covered entirely with short hairs. The wasp frequents both wetlands and drylands in search of leaffolder eggs. The wasps lay eggs in leaffolder eggs (Fig. 90) and the parasite larvae develop inside the leaffolder larva even after the host eggs hatch. The wasp eggs divide many times and 200-300 wasps are produced from a few eggs. Hundreds of wasp pupae can be seen through the skin of the leaffolder larva (Fig. 91). Figure 92 shows one of the wasps that emerged from a parasitized larva. The parasite adults live only 2-3 days.

92.1. Antenna 8-segmented
92.2. Basal one-third of front wing with few hairs

Parasite of stem borer larvae — wasp

Amauromorpha accepta metathoracica (Ashmead)

Hymenoptera: Ichneumonidae

Ichneumonids are stout to medium-sized wasps with two cross (recurrent) veins in the front wings and an elongated median cell reaching the base of the marginal vein in the hind wings. *A. a. metathoracica* is red and black with a white band at the tip of the abdomen (Fig. 93, 94). This wasp, which is common in wetlands, specializes in yellow and white stem borers. A single egg is laid in each larval host. The wasp larva emerges from the dead host and pupates inside the tunnel.

94.1. Front margin solid
94.2. Segments 2-3 entirely black
94.3. White band on segment 7

93

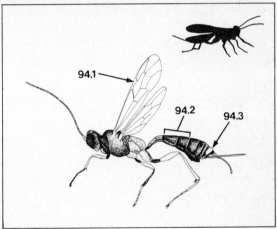

94.1

94.2

94.3

94

Index

Hymenoptera: Braconidae
 Brachymeria lasus (Walker) **96**
 Cardiochiles philippinensis Ashmead **84**
 Cotesia (=*Apanteles*) *angustibasis* (Gahan) **86**
 Cotesia (=*Apanteles*) *flavipes* Cameron **88**
 Macrocentrus philippinensis Ashmead **80**
 Opius sp. **90**
 Phanerotoma sp. **92**
 Snellius (=*Microplitis*) *manilae* (Ashmead) **94**
 Stenobracon nicevillei (Bingham) **82**
Hymenoptera: Chalcididae
 Brachymeria sp. **96**
 Brachymeria excarinata Gahan **96**
Hymenoptera: Dryinidae
 Haplogonatopus apicalis Perkins **104**
 Pseudogonatopus flavifemur Esaki and Hashimoto **104**
 Pseudogonatopus nudus Perkins **104**
Hymenoptera: Elasmidae
 Elasmus sp. **102**
Hymenoptera: Encyrtidae
 Copidosomopsis nacoleiae (Eady) **64**
Hymenoptera: Eulophidae
 Tetrastichus schoenobii Ferriere **50**
Hymenoptera: Ichneumonidae
 Amauromorpha accepta metathoracica (Ashmead) **66**
 Charops brachypterum Gupta & Maheswary **74**
 Itoplectis narangae (Ashmead) **70**
 Temelucha philippinensis (Ashmead) **78**
 Trichomma cnaphalocrosis Uchida **72**
 Xanthopimpla flavolineata Cameron **76**
Hymenoptera: Mymaridae
 Anagrus flaveolus Waterhouse **58**
 Anagrus optabilis (Perkins) **58**
 Gonatocerus spp. *56*
Hymenoptera: Pteromalidae
 Trichomalopsis apanteloctena (Crawford) **62**
Hymenoptera: Scelionidae
 Telenomous cyrus (Nixon) **54**
 Psix lacunatus Johnson and Masner **54**
 Telenomus rowani (Gahan) **52**
Hymenoptera: Trichogrammatidae
 Oligosita aesopi Girault **60**
 Oligosita naias Girault **60**
Strepsiptera: Elenchidae
 Elenchus yasumatsui Kifune and Hirashima **114**
Strepsiptera: Halictophagidae
 Halictophagus spectrus Yang **114**

MGA NAGDADALA NG SAKIT

Baculovirus: Baculoviridae
 Granulosis viruses **124**
 Nuclear polyhedrosis viruses **124**
Moniliales: Moniliaceae
 Beauveria bassiana (Balsamo) Vuillemin **118**
 Metarhizium anisopliae (Metchnikoff) Sorokin **116**
 Metarhizium flavoviride Gams and Roszypal **116**
 Nomuraea rileyi (Farlow) Samson **122**
Moniliales: Stilbaceae
 Hirsutella citriformis Speare **120**

95

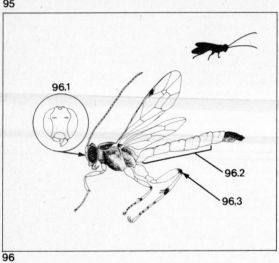

96.1

96.2

96.3

96

70

Parasite of larvae — wasp

Itoplectis narangae (Ashmead)
Hymenoptera: Ichneumonidae
Itoplectis narangae is a medium-sized wasp with a black head and thorax, orange legs, and abdomen with a black tip (Fig. 95, 96). It is a solitary hunter, searching the upper canopy mostly in wetland rice habitats. The wasp can locate larvae hidden behind leaf sheaths or inside stems. It parasitizes the larvae of leaffolder, green semilooper, hairy caterpillar, and striped and pink stem borers. A single female wasp emerges from each larval host even if the host was parasitized by several wasps. A wasp can lay 200-400 eggs over 2-3 weeks.

96.1. Eyes angular opposite antennae
96.2. Segments 1-4 or 1-5 red
96.3. Hind thigh (femur) red but black at tip

Parasite of leaffolder larvae — wasp

Trichomma cnaphalocrosis Uchida
Hymenoptera: Ichneumonidae

The large slender *Trichomma* wasp is black and yellow with an orange brown abdomen. The ovipositor of the female is half the length of the abdomen (Fig. 97, 98). When the insect is at rest, the wings are only half the length of the abdomen. Normally, the active wasps are seen flying above the leaf canopy or searching the leaves for leaffolder larvae. Older larvae are preferred. The parasite enters folded leaves to lay a single egg in each leaffolder larva encountered. A single wasp larva develops and pupates within its host, emerging from the head end of the leaffolder pupa. Sometimes the larva of *Trichomma* is parasitized by a black and yellow chalcid wasp *Brachymeria* with enlarged hind legs.

98.1. Inner margins of eyes come together in front
98.2. Wing base with yellow patches
98.3. Segment pointed
98.4. Segments 1-2 long and slender

97

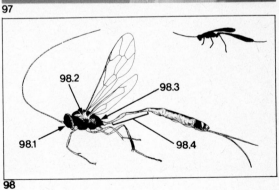

98.2

98.3

98.1

98.4

98

73

99

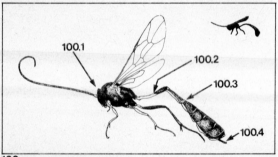

100.1

100.2

100.3

100.4

100

101

74

Parasite of larvae — wasp

Charops brachypterum Gupta and Maheswary
Hymenoptera: Ichneumonidae
The large *Charops* wasp has a black body with yellow orange markings on the bases of the antennae, legs, and abdomen (Fig. 99, 100). The abdomen is greatly enlarged at the end. The wasp searches the rice foliage for larvae of leaffolders, green semiloopers, and yellow stem borers. To parasitize the stem borer inside the stem, the wasp first locates the larva, then pierces the stem with its ovipositor and lays an egg near the host larva. The legless parasite larva hatches and then wiggles to the stem borer larva. It bites the body of the host and feeds externally on the oozing body fluids. The host larva eventually dies. When ready to pupate, the wasp larva leaves the stem and suspends itself from a leaf on a brown silk thread. It then spins a characteristically marked black and white cocoon (Fig. 101). The suspended cocoon protects the helpless pupa from predators. The adult wasp lives 3-5 days.

100.1. Eyes pointed opposite antennal base
100.2. Segment 1 very long
100.3. Abdomen slender and laterally compressed
100.4. Ovipositor short

Parasite of stem borer larvae — wasp

Xanthopimpla flavolineata Cameron
Hymenoptera: Ichneumonidae
Xanthopimpla is a medium-sized stout wasp, yellow orange in color with black markings on each abdominal segment. *X. flavolineata*, however, has no black spots on the abdomen (Fig. 102, 103). The body is coarse textured and the ovipositor is black. The wasp parasitizes stem borer larvae in dryland and wetland environments. Not a strong flier, the wasp often rests on the foliage. One wasp emerges from each stem borer pupa within the stem. After emergence, the wasps live 5-7 days.

103.1. Top of abdomen without black spots

102

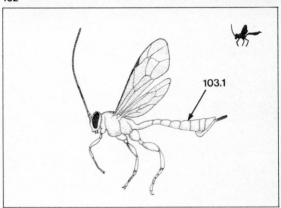

103.1

103

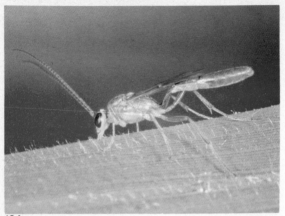

104

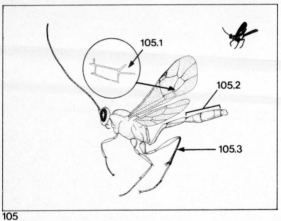

105.1

105.2

105.3

105

Parasite of larvae — wasp

Temelucha philippinensis (Ashmead)
Hymenoptera: Ichneumonidae

Temelucha is medium-sized and is a fast flier. The adult wasp is orange brown and looks like *Macrocentrus*, except that *Temelucha* has a flattened abdomen and shorter antennae (Fig. 104, 105). *Temelucha* are found in all rice environments. During the day, they hunt stem borer or leaffolder larval hosts. When stem borer larvae move from one tiller to another they are parasitized before they can rebore into the rice stem. Upon maturity, the parasite larva leaves its host and spins a light brown cocoon in the stem borer tunnel or folded leaf. The adult wasps live 7-9 days.

105.1. Tip of second rectangular (discoidal) cell blunt
105.2. Tergite 3-6 alternately black and reddish brown
105.3. Blackish brown

Parasite of leaffolder larvae — wasp

Macrocentrus philippinensis Ashmead
Hymenoptera: Braconidae

Braconids are medium to large wasps with or without a second cross (recurrent) vein. They have a short waist and a long ovipositor. The thin-bodied *M. philippinensis* is medium sized with a long orange or dark yellow abdomen. The ovipositor is twice as long as the female's abdomen and nearly as long as its antennae (Fig. 106, 107). Males are similar in size and color but lack the ovipositor (Fig. 108). *M. philippinensis* is common in all rice environments, flying above the rice canopy in search of leaffolder larvae. The single egg laid in a leaffolder larva hatches as a single parasite larva. It emerges from the leaffolder larval body and spins a brown cocoon nearby within the folded leaf.

107.1. Cell longer than wide
107.2. Abdominal segments black on top
107.3. First segment slender
107.4. Ovipositor very long

106

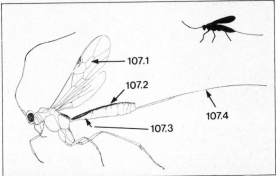

107.1
107.2
107.3
107.4

107

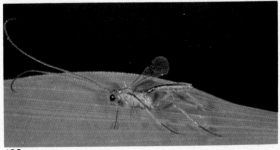

108

81

109

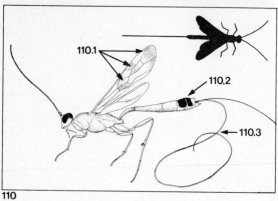

110.1
110.2
110.3

110

Parasite of stem borer larvae — wasp

Stenobracon nicevillei (Bingham)
Hymenoptera: Braconidae

The adult of *S. nicevillei* has an orange-brown body with three black markings on each front wing and two black bands on the abdomen (Fig. 109, 110). The ovipositor is twice as long as its body. This wasp is common in drylands, searching among rice plants for larvae of yellow and pink stem borers. The long ovipositor allows it to deposit a single egg into a stem borer larva inside the rice stem. One wasp emerges from each larval host.

110.1. 3 wing markings
110.2. Black bands
110.3. Very long ovipositor

Parasite of leaffolder larvae — wasp

Cardiochiles philippinensis Ashmead
Hymenoptera: Braconidae
C. philippinensis is a medium-sized stout black wasp. The tip of each wing is darkened (Fig. 111, 112). It is common in dryland and flooded rice-fields where it searches for leaffolder larvae. The wasp enters folded leaves and lays a single egg on the leaffolder larva. The creamy white wasp larva feeds externally on the host larva with a behavior similar to that of *Goniozus*.

112.1. Basal 2/3 of front wing transparent
112.2. Black

111

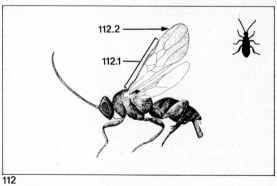

112.2

112.1

112

85

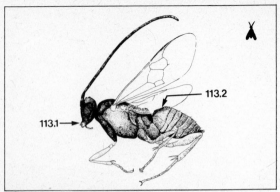

113

113.1 113.2

114

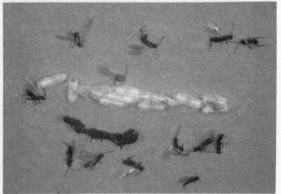

115

Parasite of leaffolder larvae — wasp

Cotesia (=*Apanteles*) *angustibasis* (Gahan)
Hymenoptera: Braconidae

There are several species of *Cotesia* in ricefields. These wasps are small but stout with clear wings. The antennae are as long as the body. The small black *C. angustibasis* wasps are commonly seen searching the rice canopy for hosts. They can be recognized by the waist-like first abdominal segment, which is three times longer than wide (Fig. 113). This wasp is found in all rice environments and specializes as a larval parasite of leaffolders. The female wasp lays more than 10 eggs inside each leaffolder larval host. The wasp larvae hatch and feed on the internal tissues of the leaffolder larva, eventually killing it. When ready to pupate, the wasp larvae leave the now dead leaffolder larva and spin white cocoons nearby (Fig. 114). The masses of white cocoons normally occur in the upper portion of leaves, outside the folded leaf, and are easy to see. Thus exposed, the pupae rely on the protective silk covering as a defense against desiccation, predators, and other parasites that would attack them (Fig. 115). Wasps live 4-5 days.

113.1. Slightly pinched
113.2. Midportion of segment 1 black, long, and slender

Parasite of larvae — wasp

Cotesia (=Apanteles) flavipes Cameron
Hymenoptera: Braconidae
C. flavipes and *C. angustibasis* are black wasps
(Fig. 116). In *C. flavipes*, however, the base of the
hind legs is brown yellow to red. *C. flavipes* is
more common in wetland rice and searches rice
plants for stem borer and green semilooper
larvae. The female lays 1-20 eggs in each host
larva. The immature parasites feed inside the
host and, in contrast to *C. angustibasis,* the larvae
emerge from the midlateral sides of the dead host
and spin overlapping cocoons nearby or below
the host (Fig. 117). All *Cotesia* larvae begin
spinning their protective white silk cocoons
before they totally emerge from their larval hosts.
Wasps live 5-7 days.

116.1. Antenna short
116.2. First segment without spots
116.3. Base at hind leg reddish and spotted

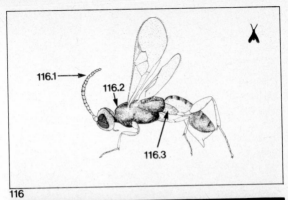

116

117

89

118

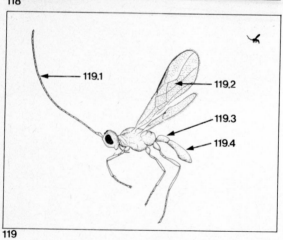

119.1

119.2

119.3

119.4

119

Parasite of whorl maggot larvae — wasp

Opius sp.
Hymenoptera: Braconidae
Opius are small, orange-brown wasps with long antennae (Fig. 118, 119). They parasitize whorl maggot larvae. One parasite larva develops in each larval host. *Opius* wasps emerge in 7-9 days from the whorl maggot pupae, and live 3-4 days. *Opius* larvae in turn may be parasitized by *Tetrastichus*, a small black wasp which also emerges from the whorl maggot pupa.

119.1. Long antenna
119.2. 4-sided
119.3. Black and slightly triangular on top
119.4. Finely spotted

Parasite of stem borer larvae — wasp

Phanerotoma sp.
Hymenoptera: Braconidae

Phanerotoma is a small, light brown wasp which has three segments on its broad but short abdomen (Fig. 120). The wasp (Fig. 121) parasitizes stem borers by laying one egg in each larva. The parasite larva develops within the body of the stem borer until the stem borer pupates. Adult parasites emerge in 2-6 days from the stem borer pupae. Wasps live 3-5 days.

120.1. Antennal segments smaller at the end
120.2. Slightly depressed abdomen with 3 visible segments

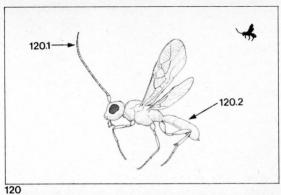

120

121

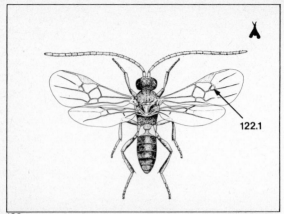

122

122.1

123

Parasite of cutworm larvae — wasp

Snellenius (=*Microplitis*) *manilae* (Ashmead)
Hymenoptera: Braconidae
Snellenius (Fig. 122), which can be confused with *Cotesia* in appearance, has a small closed cell on the fore wing and has hairy eyes. It prefers dryland habitats and specializes in parasitizing cutworm larvae. A female lays 3-5 eggs in the body of a cutworm larva. The parasite larvae consume the cutworm and spin a brownish cocoon nearby (Fig. 123). Wasps emerge 4-8 days later and live 6-8 days. *Snellenius* larvae in turn can be parasitized by *Brachymeria*, a black and yellow wasp with enlarged hind legs.

122.1. Small triangular closed cell

Parasites of larvae/pupae — wasps

Brachymeria lasus (Walker)
Brachymeria excarinata Gahan
Brachymeria sp.
Hymenoptera: Chalcididae

The three black, globular *Brachymeria* can be distinguished by yellow markings on the legs. *B. lasus* (= *obscurata*) has a triangular cheek and yellow marks on the tip of the femur and ventral half of the tibia. *B. excarinata* has no triangular cheek but has yellow markings on both ends of the black hind tibia (Fig. 124, 125). The other *Brachymeria* sp. has no yellow markings (Fig. 126). The three parasitize older larvae of leaf-folders, skippers, and satyrids. A single egg is deposited in the body of each larval or pupal host and pupation occurs inside the host. Adult wasps emerge from the head portion of the pupa and live 3-5 days.

125.1. Prominently spotted
125.2. Long first plate (tergite)
125.3. Enlarged thigh (femur)

124

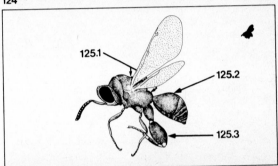

125

126

97

127

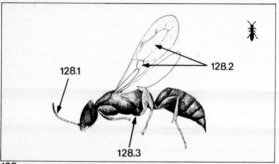

128.1

128.2

128.3

128

129

Parasite of leaffolder larvae — wasp

Goniozus nr. *triangulifer* Kieffer
Hymenoptera: Bethylidae

Bethylids are small ant-like wasps with flat bodies and few wing veins. *Goniozus* sp. has a black body and transparent wings (Fig. 127, 128). The wasp crawls on the rice leaves, searching for leaffolder larvae in wetland and dryland habitats. The wasp enters the folded leaf and paralyzes the host larva (Fig. 129) before laying 3-8 eggs outside its body.

128.1. Antennae attached below front of eyes
128.2. Cells of front wing with undefined margins
128.3. Body depressed and ant-like

Goniozus nr. *triangulifer* Kieffer

The early-stage parasite larvae are globular and yellow and feed externally on the leaffolder larva (Fig. 130). Now becoming ovoid, the parasite larvae kill their host by the fourth day (Fig. 131). Pupation occurs on the fifth day beside the leaffolder body (Fig. 132) as reddish brown cocoons form (Fig. 133). Parasite development from egg to adult stage takes 10-14 days. The female wasp lives 2-3 weeks.

130

131

132

133

134

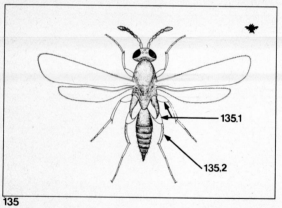

135.1

135.2

135

Parasite of leaffolder larvae — wasp

Elasmus sp.

Hymenoptera: Elasmidae

Elasmids are elongate small wasps with pointed abdomens. The segment attaching the legs to the body is enlarged and disc-shaped. *Elasmus* is black with a black and red abdomen (Fig. 134, 135). It occurs in all rice environments and parasitizes leaffolder larvae. One or two eggs are laid in each young or old larva. *Elasmus* larvae are highly aggressive and will kill other parasite larvae which may be developing inside the leaffolder host. The adults emerge either from the larva or pupa and live 2-4 days.

135.1. Thin and broad thighs (femurs)
135.2. Diamond-shaped band

Parasites or predators of hoppers — wasps

Haplogonatopus apicalis Perkins
Pseudogonatopus nudus Perkins
Pseudogonatopus flavifemur Esaki and Hashimoto
Hymenoptera: Dryinidae

Dryinid wasps are ant-like in appearance. Females are usually wingless and have a pair of pincher-like front claws to grasp prey. Males are winged. *Haplogonatopus* and *Pseudogonatopus* are common in wetland ricefields. *Haplogonatopus* attack leafhoppers (Fig. 136) and *Pseudogonatopus*, planthoppers (Fig. 137). *P. nudus* has a brown body (Fig. 137, 138) and *P. flavifemur* is black (Fig. 139).

138.1. Ant-like body brown
138.2. Pincher claw

136

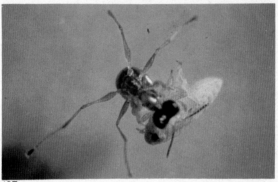

137

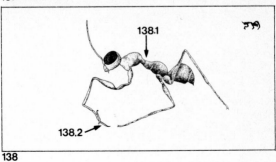

138

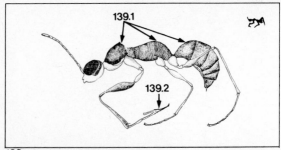

139

140

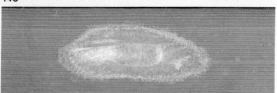

141

106

Haplogonatopus apicalis Perkins
Pseudogonatopus nudus Perkins
Pseudogonatopus flavifemur Esaki and Hashi-moto

A female lives 6-7 days and acts as a predator by attacking 2-4 prey a day. During the same period, 4-6 hosts may also be parasitized.

One to two eggs are laid singly inside the hopper host. In 1-2 days, the eggs hatch and the larvae feed on the body fluids. As they grow, the parasite larvae, each covered with a black to grayish sac, protrude from the abdomen of their hosts (Fig. 140). In 7-10 days, the sac splits and the whitish larva wiggles free. The larva then pupates and secretes a whitish oval silk cocoon to cover the pupa on the plant (Fig. 141). The flat cocoon turns reddish brown when the wasp is ready to emerge.

139.1. Uniformly black body
139.2. Pincher claw

Parasites of leafhoppers — big-headed flies

Tomosvaryella subvirescens (Loew)
Tomosvaryella oryzaetora (Koizumi)
Diptera: Pipunculidae
Pipunculids are small black flies with large round heads formed entirely by the compound eyes. *T. oryzaetora* has black shoulders, a brownish tinge on the front wings, and no hairs at the bases of hind femurs while *T. subvirescens* (Fig. 142, 143) has yellow shoulders, transparent wings, and a hairy ridge on each femur.The pupae of *T. oryzaetora* are dark red (Fig. 144).

Big-headed flies alight on the back of leafhoppers and oviposit into the host's abdomen. A single fly develops from each leafhopper. After developing inside its host, it pupates in the soil or at the base of the plant. The egg develops into an adult in 30-40 days. Flies live 4 days and parasitize 2-3 hoppers per day.

143.1. Black spot (stigma) absent
143.2. Yellow shoulder
143.3. Female ovipositor straight
143.4. Hind basal leg segment (trochanter) flat
143.5. Posterior end of male's abdomen slightly twisted on the right side

144.1. Big eyes
144.2. Black shoulder
144.3. Wings with brownish tinge

142

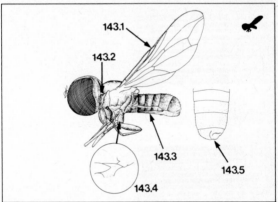

143

143.1
143.2
143.3
143.4
143.5

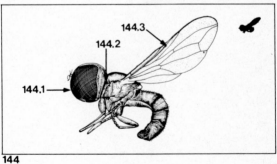

144

144.1
144.2
144.3

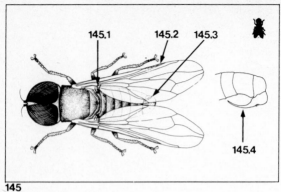

145

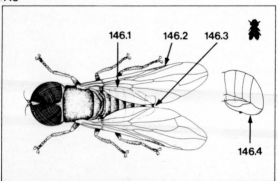

146

147

Parasites of leafhoppers — big-headed flies

Pipunculus mutillatus (Loew)
Pipunculus javanensis de Meijere
Diptera: Pipunculidae

On the other hand, *P. mutillatus* (Fig. 145) has yellowish brown tibia and tarsi, and prominent longitudinal groove in the male genitalia while *P. javanensis* (Fig. 146) has dark brown markings, and smooth male genitalia. *P. javanensis* has been reared from *Deltocephalus* leafhopper (Fig. 147).

145.1. Row of hairs
145.2. Brown spot
145.3. Tip of abdomen divided on the right side
145.4. Ovipositor of female curved toward the basal three abdominal segments

146.1. Row of hairs
146.2. Brown spot
146.3. Tip of abdomen indented at center
146.4. Ovipositor of female curved toward the basal five abdominal segments

Parasites of hoppers — strepsiptera

Halictophagus spectrus Yang
Strepsiptera: Halictophagidae
Elenchus yasumatsui Kifune and Hirashima
Strepsiptera: Elenchidae

The Strepsiptera are minute, twisted-winged parasites related to beetles. The wingless female remains inside the host and only its head protrudes from the abdominal segments (Fig. 148). Adult males have knob-like front wings and membranous fan-shaped hind wings. They fly to the parasitized host to mate with the sedentary female. A fertilized female produces 500-2,000 larvae, which crawl out of the moribund host to search for new hosts. The larvae use their well-developed eyes, legs, and sensory body hair to locate hosts. They bite into a host with their mandibles and penetrate body membranes for successful parasitization. *Halictophagus* is specific to leafhoppers and *Elenchus* to planthoppers. *H. spectrus* (Fig. 149, 150) males are black, have 3-segmented tarsi, and 7-segmented broad antennae while *E. yasumatsui* (Fig. 151) are light brown, have 2-segmented tarsi, and 4-segmented antennae. *E. yasumatsui* triungulids are minute and C-shaped with black head and pale brown body (Fig. 152). Males live 1-2 days, females live 1-2 months.

150.1. Antenna 7-segmented
150.2. Transparent wings with more veins
150.3. Foot (tarsus) 3-segmented

151.1. Antenna 4-segmented
151.2. Foot (tarsus) 2-segmented
151.3. Clubbed fore wing
151.4. Wings transparent with few veins

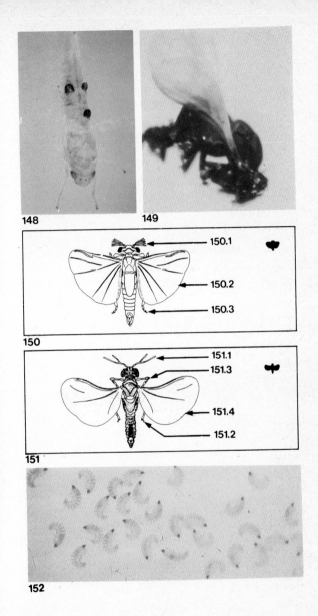

148

149

150
150.1
150.2
150.3

151
151.1
151.3
151.4
151.2

152

153

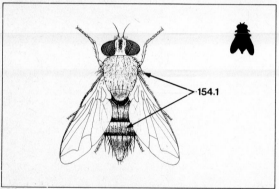

154.1

154

155

Parasite of skipper larvae — fly

Argyrophylax nigrotibialis (Baranov)
Diptera: Tachinidae

Tachinids are gray or black spiny flies slightly larger than a housefly (Fig. 153, 154). *A. nigrotibialis* parasitizes skipper larvae in wetland and dryland environments. The female fly hovers over the rice foliage in search of larval hosts. When a host is found, she lands on the back of the larva and places 2-4 eggs on its body. The hatching maggots rasp open a wound in the host and enter the host's body. The fully fed maggots secrete a hard cocoon or puparium, changing from light yellow to dark red. The puparia, near the dead skipper larva, are covered with a white powder (Fig. 155). Adult flies emerge in 4 days and live for 3 more days to mate and seek new skipper larvae.

154.1. Thorax and abdomen with many long hairs

Pathogen — fungus diseases

Metarhizium anisopliae (Metchnikoff) Sorokin
Metarhizium flavoviride Gams and Roszypal
Moniliales: Moniliaceae

Metarhizium fungi infect hoppers, bugs, and beetles. The spores land on the body of an insect and, under conditions of prolonged high humidity, germinate and grow into the insect body. The fungus growing within the insect host consumes the body contents. When the insect dies, the fungus emerges first as white growth from the host body joints as shown here on the black bug (Fig. 156). When spores are formed, the fungus turns dark green if it is *M. anisopliae* (Fig. 157) or light green if it is *M. flavoviride*. The latter has parasitized a zigzag leafhopper (Fig. 158). Spores emerging from the dead host are spread to new hosts by wind or water.

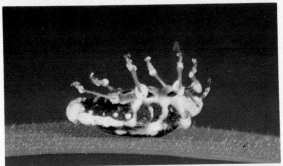

156

157

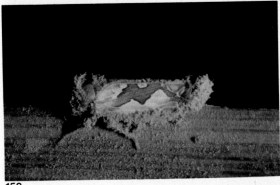

158

117

159

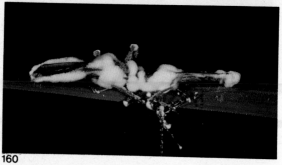

160

Pathogen — fungus disease

Beauveria bassiana (Balsamo) Vuillemin
Moniliales: Moniliaceae
Beauveria bassiana is a white fungus which attacks planthoppers, leafhoppers, stem borers, leaffolders, rice bugs, and black bugs. It occurs in all rice environments and, like other fungal diseases, requires conditions of prolonged high moisture for the airborne or waterborne spores to germinate. The fungus invades the soft tissues and body fluids of its host, and grows out of the body when ready to produce the dispersing spores. The spores appear chalky white as on the body of the brown planthopper (Fig. 159), or rice bug (Fig. 160).

Pathogen — fungus disease

Hirsutella citriformis Speare
Moniliales: Stilbaceae
Hirsutella is a fungus that infects planthoppers and leafhoppers. After the fungus enters the body of the host and consumes its inner tissues, it grows out as long filaments that are dirty white at first (Fig. 161) and later turn grey (Fig. 162). The dispersing infectious spores are produced from the filaments.

161

162

121

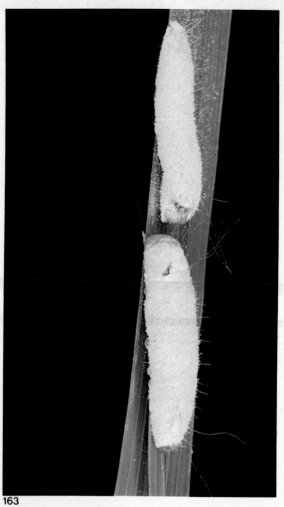

163

Pathogen — fungus disease

Nomuraea rileyi (Farlow) Samson
Moniliales: Moniliaceae
Nomuraea is a white fungus with pale green spores (Fig. 163). It attacks the larvae of stem borers, leaffolders, green hairy caterpillars, armyworms, and caseworms.

Early in the infective stages, larvae attacked by *Nomuraea* become white. After a few days, spores are formed and the caterpillars become pale green.

Pathogen — virus diseases

Nuclear polyhedrosis viruses
Baculovirus: Baculoviridae
Nuclear polyhedrosis viruses are common on armyworms and cutworms. The larvae become infected by eating virus contaminated foliage. When the virus spreads in the larva's body, the host becomes sluggish and stops feeding (Fig. 164). Later the larva turns whitish and then black and hangs from the rice foliage by the larval body prolegs (Fig. 165). The oozing fluid from the larval body contaminates nearby foliage and continues the disease cycle.

Granulosis viruses
Baculovirus: Baculoviridae
Granulosis viruses attack moth and butterfly larvae. As with the nuclear polyhedrosis virus, the host larva that eats contaminated foliage slows its movement and later stops feeding. After 1 to 2 weeks, the body becomes constricted, giving a segmented appearance as in the larva of the brown semilooper (Fig. 166). Infected larvae turn yellow, pink, and black. Virus-infected larvae become soft.

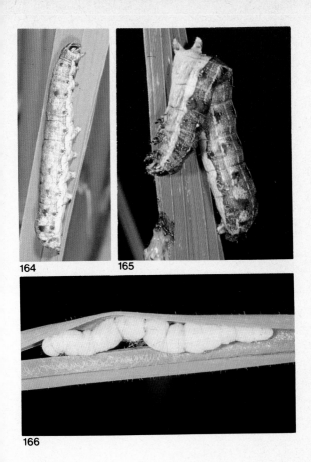

164

165

166

125

Acknowledgments

We thank Mr. Semy Lapiz who took the pictures for Figures 38, 75, 83, 84, 86, 117, 136, 140, 142, 147, 151, and 152; and FAO which supplied the film. We acknowledge the assistance of Mr. Danilo Amalin for the drawings and Mrs. Liberty Almazan for her help with photographing the specimens.

The International Rice Research Institute (IRRI) was established in 1960 by the Ford and Rockefeller Foundations with the help and approval of the Government of the Philippines. Today IRRI is one of the 13 nonprofit international research and training centers supported by the Consultative Group on International Agricultural Research (CGIAR). The CGIAR is sponsored by the Food and Agriculture Organization (FAO) of the United Nations, the International Bank for Reconstruction and Development (World Bank), and the United Nations Development Programme (UNDP). The CGIAR consists of 50 donor countries, international and regional organizations, and private foundations.

IRRI receives support, through the CGIAR, from a number of donors including: the Asian Development Bank, the European Economic Community, the Ford Foundation, the International Development Research Centre, the International Fund for Agricultural Development, the OPEC Special Fund, the Rockefeller Foundation, the United Nations Development Programme, the World Bank, and the international aid agencies of the following governments: Australia, Canada, China, Denmark, France, Federal Republic of Germany, India, Italy, Japan, Mexico, Netherlands, New Zealand, Norway, Philippines, Saudi Arabia, Spain, Sweden, Switzerland, United Kingdom, and United States.

The responsibility for this publication rests with the International Rice Research Institute.

Notes

Notes

Notes

Notes

Notes

Notes

Notes

Notes

Notes